Living Green

Saving Energy

By Meg Gaertner

level 2 little blue readers

www.littlebluehousebooks.com

Little Blue House is distributed by North Star Editions:
sales@northstareditions.com | 888-417-0195

Produced for Little Blue House by Red Line Editorial.

Photographs ©: Shutterstock Images, cover, 4 (top), 7, 8, 10–11, 13 (top), 13 (bottom), 15, 18 (top), 18 (bottom), 21, 22–23, 24 (top left), 24 (top right), 24 (bottom left), 24 (bottom right); iStockphoto, 4 (bottom), 17

Library of Congress Control Number: 2022901878

ISBN
978-1-64619-600-5 (hardcover)
978-1-64619-627-2 (paperback)
978-1-64619-678-4 (ebook pdf)
978-1-64619-654-8 (hosted ebook)

Printed in the United States of America
Mankato, MN
082022

About the Author

Meg Gaertner enjoys reading, writing, dancing, and being outside. She lives in Minnesota.

Table of Contents

gases

Why Save Energy?

Burning fuel makes energy. But burning fuel also sends harmful gases into the air. These gases are dirty and hurt people's health.

By saving energy, people can burn less fuel. Burning less fuel will keep the air clean and keep people healthy.

Using Less Energy

Using less energy is one way to save energy.

There are many simple ways to use less energy.

Instead of turning on lights, people can open curtains for sunlight. They can also turn off lights when they leave rooms.

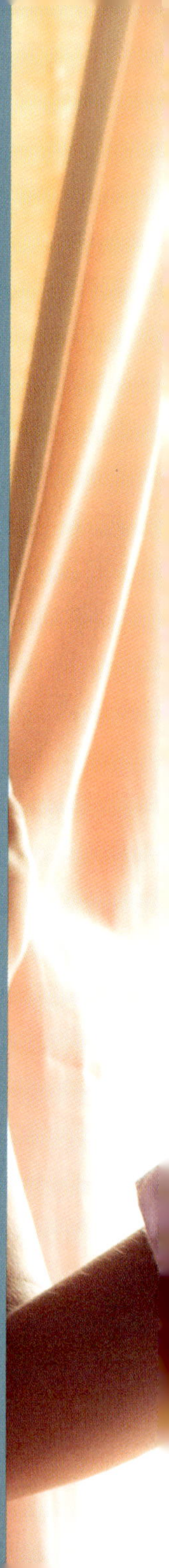

curtain

Electronic devices use energy when they are plugged in.
So, people can unplug electronic devices to save energy.

electronic devices

plug

Cars burn fuel to make energy. That makes them move. People can use less energy by walking or biking instead.

biking

Making new items also uses lots of energy. People can save energy by buying less. They can reuse old items instead.

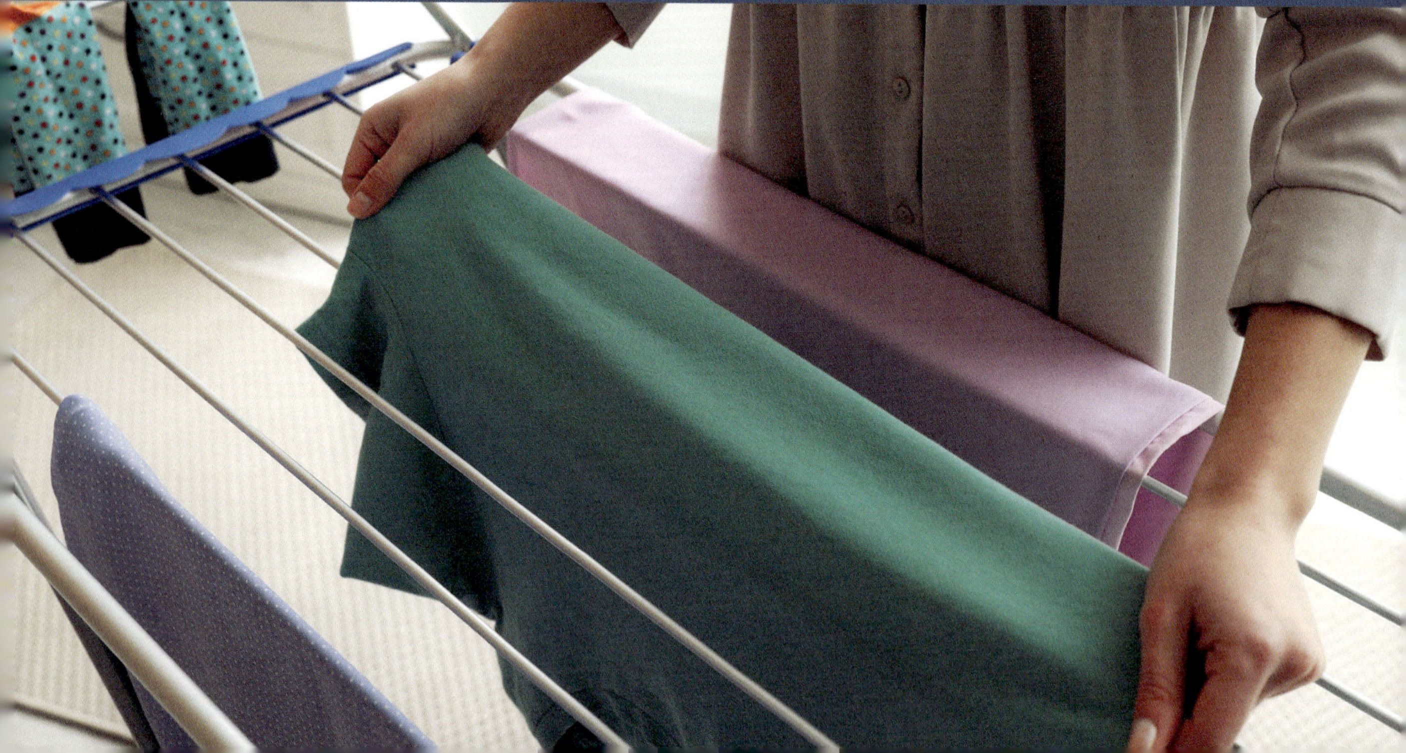

Energy Efficiency

Energy efficiency also helps save energy.
Energy efficiency means using less energy for the same job.

For example, all light bulbs do the same job. But LED light bulbs use less energy. Changing to LED light bulbs saves energy.

LED light bulb

Burning fuel is one way to get energy.

But using wind, water, or sunlight is cleaner.

Switching to clean energy can help save Earth.

Glossary

clean energy

fuel

electronic devices

LED light bulb

Index